Build Your Own Free To Air Antenna Solution

By Ryan Seager

***

Build Your Own Free To Air Antenna Solution

By Ryan Seager

***

Table of Contents

Introduction

This book will show you how to build a powerful Free-To-Air Antenna solution that will allow you to get free HD channels on your television. I live about 60 miles from a major metropolitan city and these designs allow me to pick up several channels with perfect clarity. If you live within a major metropolitan city, you can use these designs to build an antenna that will give you ten, twenty, thirty channels or more depending on the city.

Obviously the farther you are from television broadcasting stations, the fewer channels you will be able to get. That's why this is called free-to-air. The television towers broadcast the signals out into the air and you can access them for free if you have the right equipment and if you are within range.

It is very likely that you have an HDTV or a 'HDTV-Ready' set, and you want the best way to get free local HD channels. You may be fed with the monthly expense of paying for cable when all you need is the right antenna hooked up correctly.

Almost all TV stations no longer broadcast in the UHF/VHF spectrum. Those were the old stations you got that were normally from channel 2 to channel 13 on the 'hand dial' TV sets (this was before we got the wonderful technology of the remote control and you actually had to turn the dial by hand!). Those stations would come in fuzzy depending on how well you could position the 'rabbit ears' or the exterior TV antenna on the roof. Nowadays the channels are broadcast in high definition which gives either a very clear picture or no picture at all. (Sometimes you will get a pixilated image if the signal is weak, but in general it is super clear or nothing.)

Probably if you look at the back of your television, you will see the following connections:

The input on the right is where you attach your regular cable input line (RG 62 wire) from the cable company. This is the connection that costs you every month from your local cable service provider.

Now if you look to the left in the above image, the connection labeled ANT 1 IN (Air); that is what we are interested in! Our goal is to remove the connection on the right, the ANT 2 IN (cable), and replace it with a connection that goes to the ANT 1 IN.

The connector that we will attach to the ANT 1 IN connection will NOT be a standard cable wire. Instead we will attach a transformer connector that looks like this:

Or a quick snap-on connector like this:

These are called '75 to 300 OHM Transformers' or sometimes called 'Matching Transformers'. You can pick them up for about 5 bucks from retailers like 'Radio Shack' or 'The Source' or any local electronic parts supplier.

The round end goes to the ANT 1 IN connection on your HD ready TV and the two leads attach to regular 300 OHM antenna wire. You will simply attach the regular antenna wire to the HD antenna you will build to get the free high quality Digital signals; but we are getting a bit ahead of ourselves. Let's take this step by step so that you can see clearly the simple process in detail.

Chapter 1 – Materials List

Here is a list of things you will need to build your own Free to Air antenna:

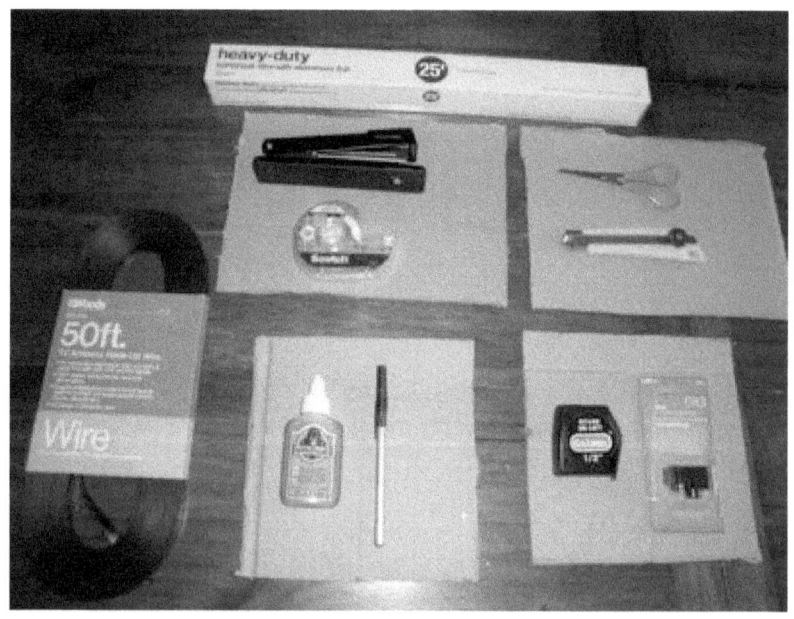

Aluminum Foil – Preferable 'heavy duty' type.

Cardboard – Cut into four pieces, 2 pieces of 8" x 8" and 2 pieces of 11"x 8"

300 Ohm Antenna Wire

Stapler –

Cellulose Tape

Scissors-

Razor Knife or Exact-O-Knife or Box Cutter Knife –

Glue-

Measuring Tape-

Pen or Marker-

75 Ohm to 300 Ohm Transformer-

Chapter 2 – Building the Antenna

With the above materials, you can create many different Antenna styles in different shapes. For our first design we will make one that is called 'Log-Periodic' or Logarithmic style. This is named for the mathematical design of the following template.

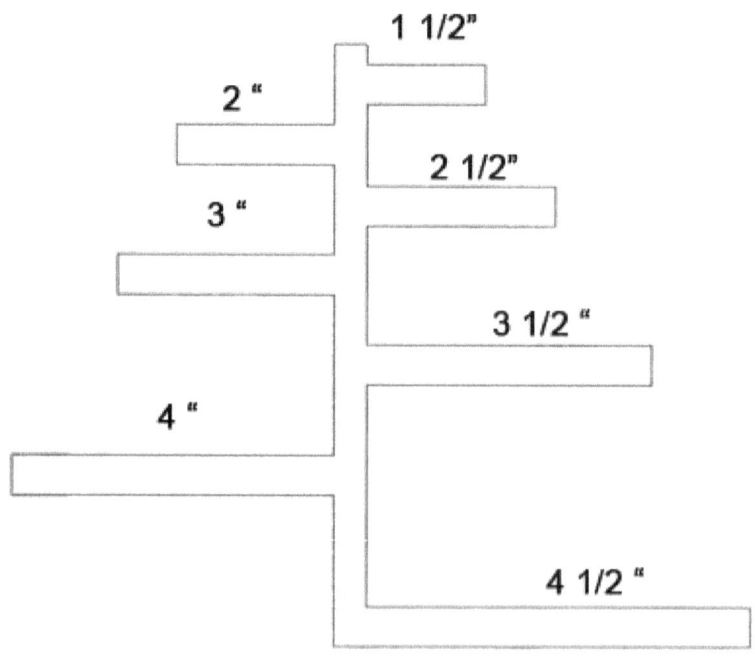

The arms of this antenna 'tree' are a half inch wide and then the lengths are shown in the image. The first arm is 1 ½ inches long, the second arm is 2 inches long and so on. It may be a bit tricky but with a simple piece of paper, a pen and a tape measure or ruler, you can quickly draw this out on a standard 8 ½ by 11 sheet of printer paper. (turn the paper sideways, that is landscape orientation, to fit the dimensions of the antenna template).

Once you have drawn the template, cut it out with a pair of scissors as in the image below:

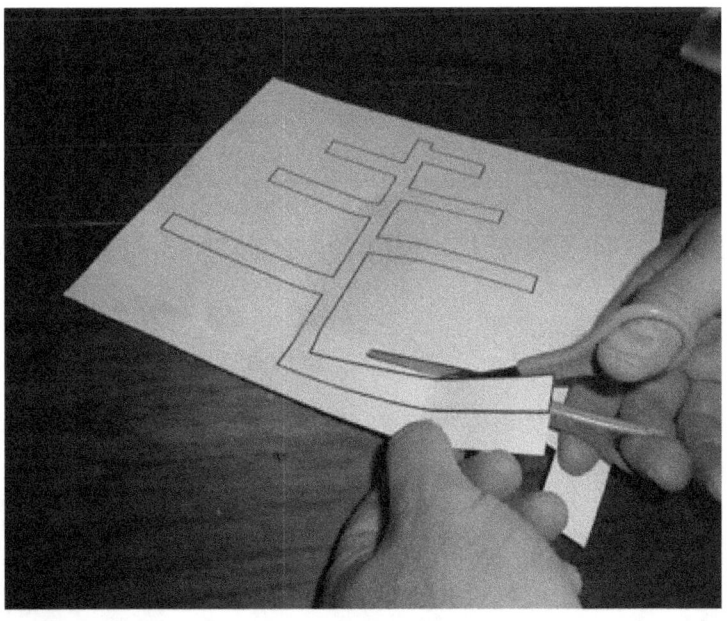

Once you have cut out your template, you will place it on a sheet of aluminum foil as in the image below:

You may want to temporarily tape the template to the foil so that it does not move. Then mark an outline around the template using a pen or marker so that it appears on the aluminum foil as in the image below:

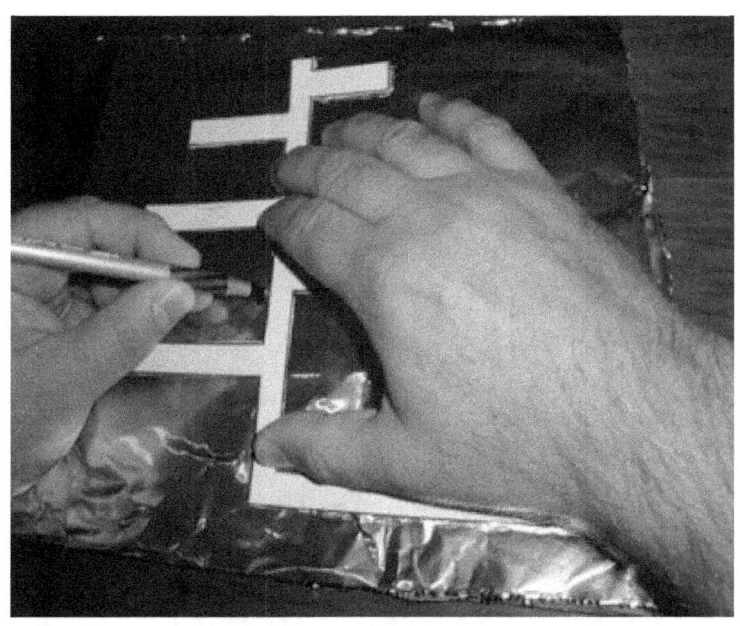

Once you are done, you will have an outline of the template on the foil like so:

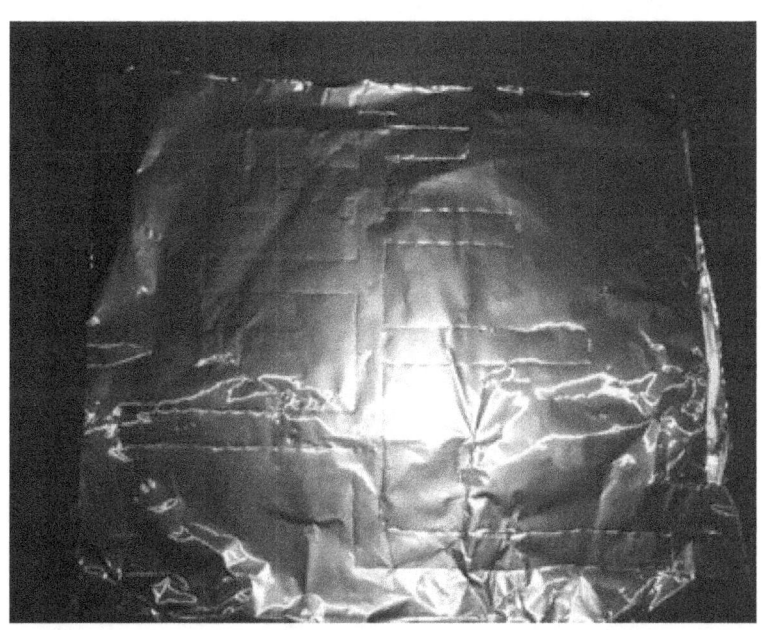

Now take your pair of scissors and cut out the template shape from the foil. This is the main antenna shape and you will need to do this twice to create two halves of the antenna design.

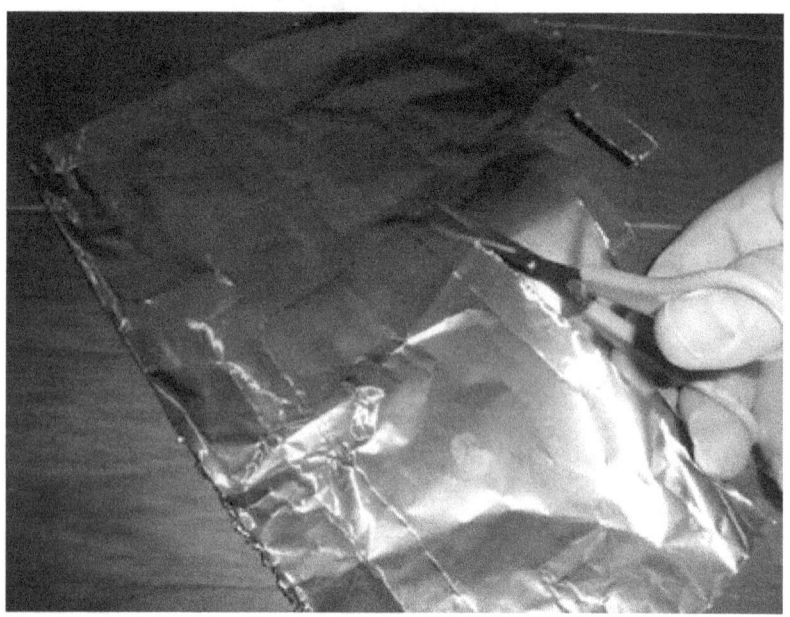

Once you have cut out the template shape from the aluminum foil, take one of the pieces of 11 x 8 inch cardboard and find the center.

Lay a bead of glue along this center line like so:

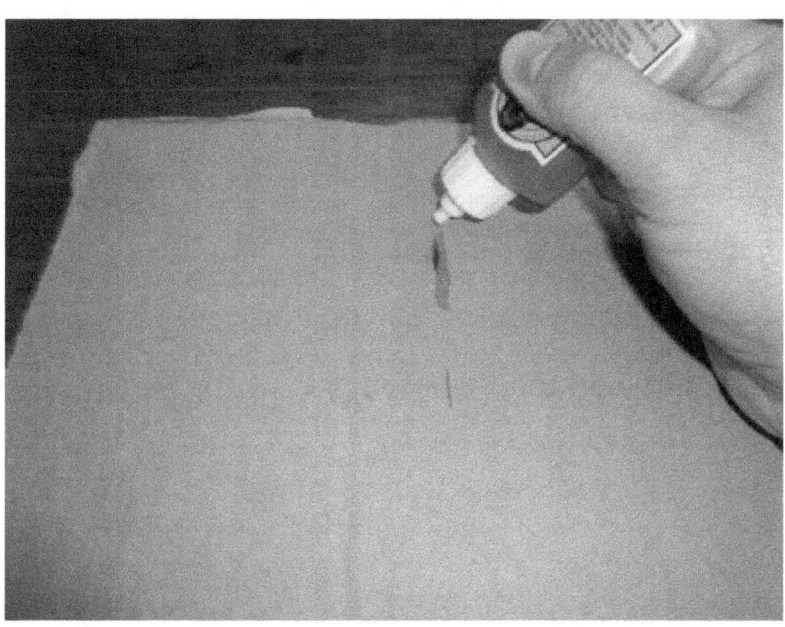

Now place the Aluminum 'antenna' template onto the cardboard.

Now you will glue it to the cardboard. You can see in the image, I am using Gorilla Grip brand of glue. But you can also use any craft glue, preferable something that is non-toxic and easy to clean up if you get some on your hands or workspace.

Once you have glued the aluminum foil to the cardboard, you are ready for the next step. Take the 300 Ohm antenna wire and carefully separate the two wires and strip off the last two inches of plastic covering.

Antenna wire is too small for most wire stripper tools so you will be better off to use a razor-knife like from our material list:

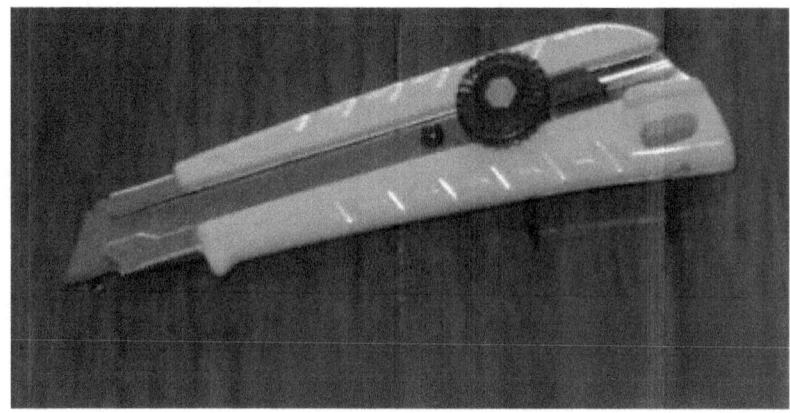

Razor Knife

Once you have stripped the antenna wire, you want to carefully position one of wires onto the narrow end of your 'log-tree' antenna like in the image below:

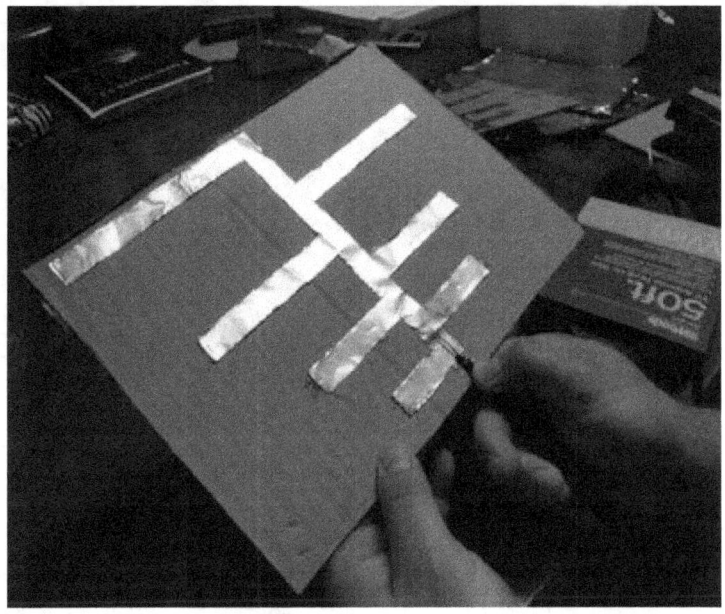

This antenna design has two-halves so you will staple one wire to each half. The staple will go through the foil and the cardboard.

You can use two staples as in the image below:

Now that you have the two halves of the antenna, you will take them and position into a V-shape (the angle is about 45 degrees but it does not have to be exact) onto the two pieces of cardboard that measure 8" x 8" as in the image below:

This creates a type of frame for the V-Shaped two 8" x 11" sections. If you were look into the middle of the antenna, it would look like this:

You can keep the whole antenna unit together with tape and then glue the cardboard pieces together so it will remain firmly connected. It may get a bit messy with glue but don't worry, it will not hurt the performance. Lastly you will connect the other end of the antenna wire to your 75 to 300 Ohm transformer so that you can hook it up to the ANT-1 IN connection on the back of your television.

Your antenna is now complete and you can face the V section towards the general direction of your local TV broadcast stations.

You can experiment with the position of your antenna as each location is of course unique; remember that the higher you place the antenna, the more effective will be the reception. I used 50 feet of antenna wire and ran it up to the attic and got excellent reception. You could position it outside but it would have to be covered to protect it from being destroyed by the weather; better would be to keep it inside, preferably in an attic or facing an upper level window.

Chapter 3 – Comparison Results

I came up with the above design after studying a ton of academic articles about antenna theory, radio frequency transmission and the electromagnetic spectrum; a lot of it was very interesting and some of it was quite boring. I hope the design I have provided saves you some time of unnecessary study. After having good success with the above design, I decided I would like to see how it compared to a store-bought model.

I bought an inexpensive $25 antenna from a local retailer to compare the reception with the above model I had built myself. The model I bought looked like this:

It advertised on the box that it would get HDTV channels. The unit itself is very simple and simply needs to be plugged in.

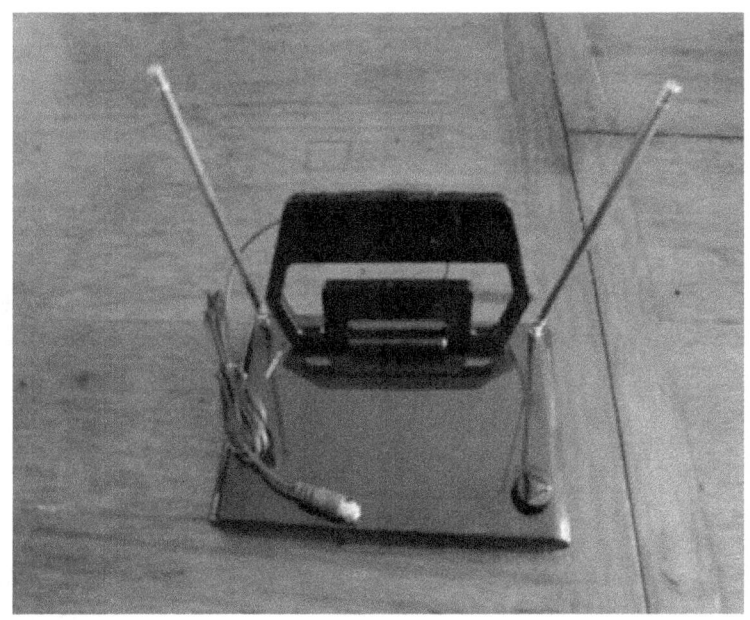

When I plugged it in and tried it out, it only got one channel. It would only work if the antenna was above the television.

Once I place the unit beside the television on the same level as the television, the picture pixilated and cut out as in the image below:

So it was clear that my home-made design was better than this basic store-bought model.

Chapter 4 – In Case You Would Rather Buy

Having seen how to make your own Free-to-Air antenna, one may wonder is it worth the effort? It is certainly a fun project but I can understand that some people, after seeing the steps involved, may not want to go through the hassle of making their own and would rather just buy a free-to-air antenna.

For those people I thought I should include some recommendations that may be helpful if you do decide to purchase.

I searched through a ton of products, talked with people who actually used them and tested a number of models myself. I am lucky enough to be friends with the owner of a satellite supply company and he let me test a number of products. After all my research I came to the conclusion of what I would buy if I decided not to make one myself.

I would buy the Eaglestar Pro Model Ant 53-6163 for a retail price of about $50 bucks. The package looks like this:

Front

Back.

This model is easy to setup, works indoors or out and gives as good reception as some models costing $100 or more. It is small and compact and seems to have good omni-directional capability so that you don't have to keep adjusting it to get better reception.

I did not find this model on eBay or Amazon so it is certainly not the most popular model out there. Please keep in mind that there are new products coming out all the time so you can be sure that they will keep getting better and better. I can only tell you the best value I found based on cost versus performance. There are definitely more complicated solutions that use bigger antennas with signal boosting devices and so on. I just thought it would be helpful to include this model since I tested the performance first hand of over a dozen products and this one came out on top.

If you do decide to purchase a Free-to-Air solution, made sure that you have an option to return for a refund since each product may perform differently depending on the unique situation of your particular home. You don't want to get stuck with something that works on display at the retail outlet but then doesn't work at your home.

Chapter 5 – An Alternative Design

Another popular design is called the 'cookie tin' design; so named, of course, because you make it from a cookie tin. I tried this design but because my accessible windows do not face towards the nearest broadcast tower, I had a hard time getting good signals. However some people swear by it so I am including the design here in case it works for you.

Step one: Get the top off your favorite cookie tin!

Side view of cookie tin.

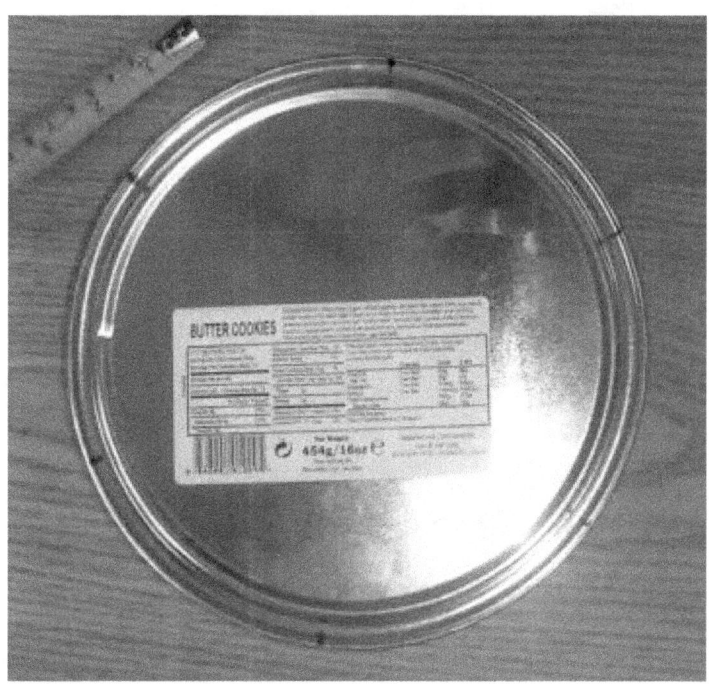

Turn over the bottom section of the tin and divide the circumference into six evenly spaced sections and mark the sections with a marker.

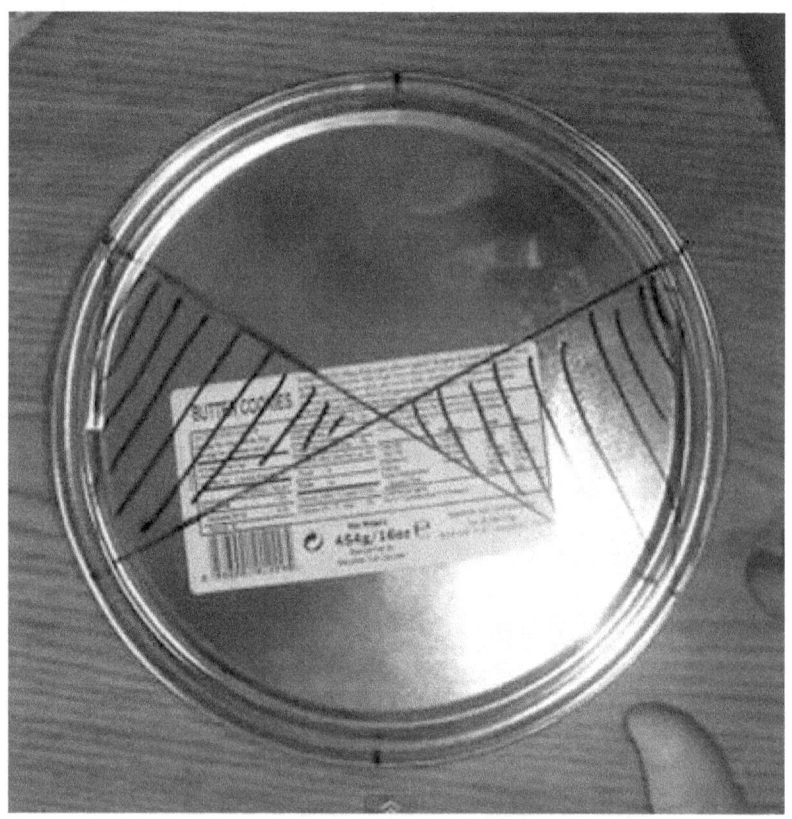

Mark off two opposites sections like in the above diagram. These sections will be cut out.

You can use a dremel to out out the sections from the cookie tin lid. You can also use tin snips.

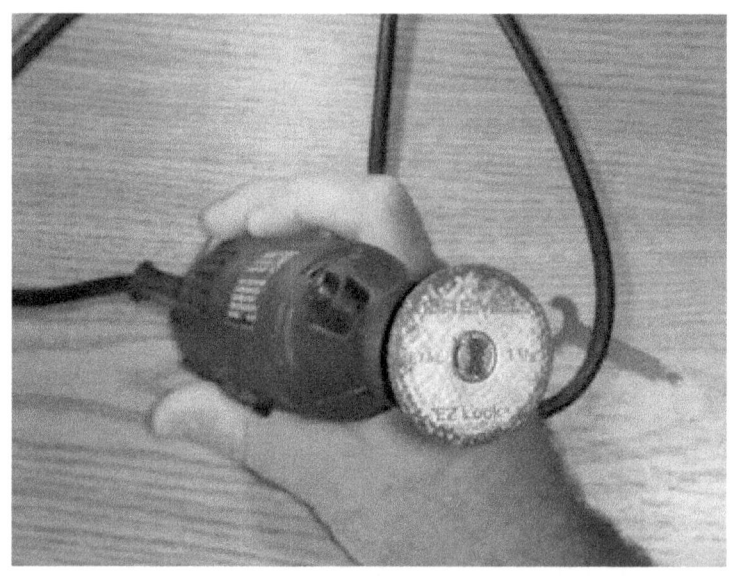

Attach a cutting blade to the dremel and remember to use eye protection.

Cookie tin with sections cut out.

Round off the edges of the sections and leave a ¼

inch gap between the cut edges.

Drill 1 ¼ inch hole at one edge of the tin where you

will run your coax cable.

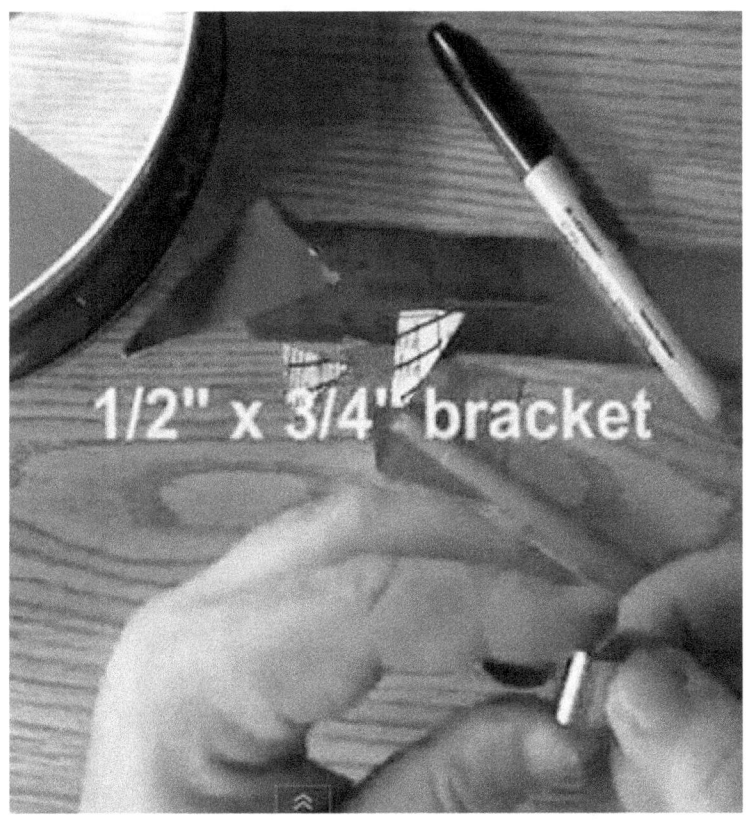

Form a small bracket from the leftover cutout pieces of the cookie tin. This step is optional since you could also use wire or strong tape. This bracket will be used to help hold the coax to the antenna.

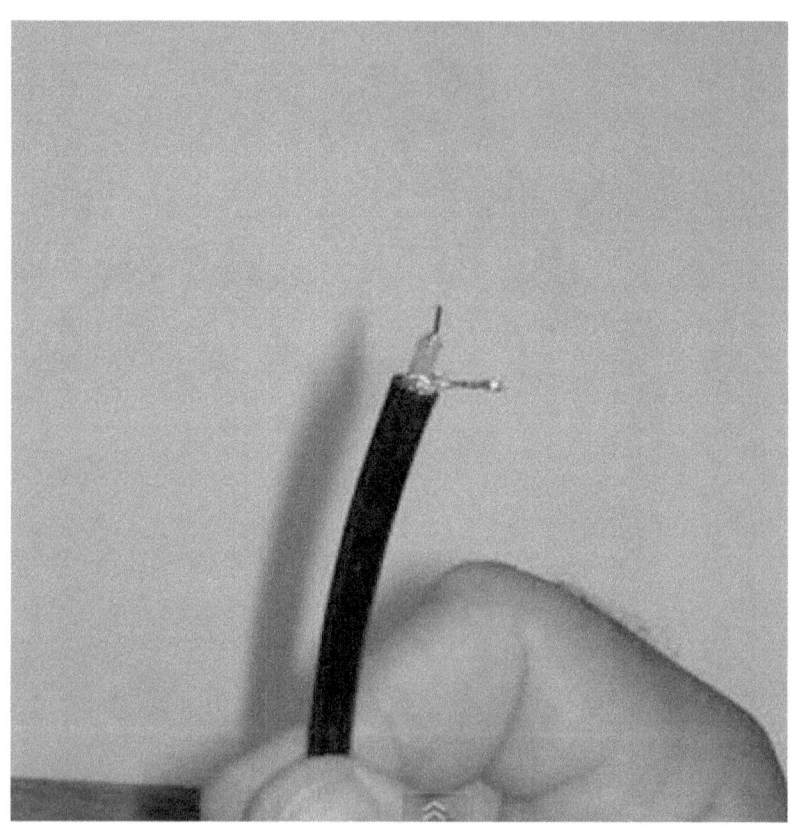

Prepare the coax (RG-62) by stripping away a portion of the insulating layers from the central wire.

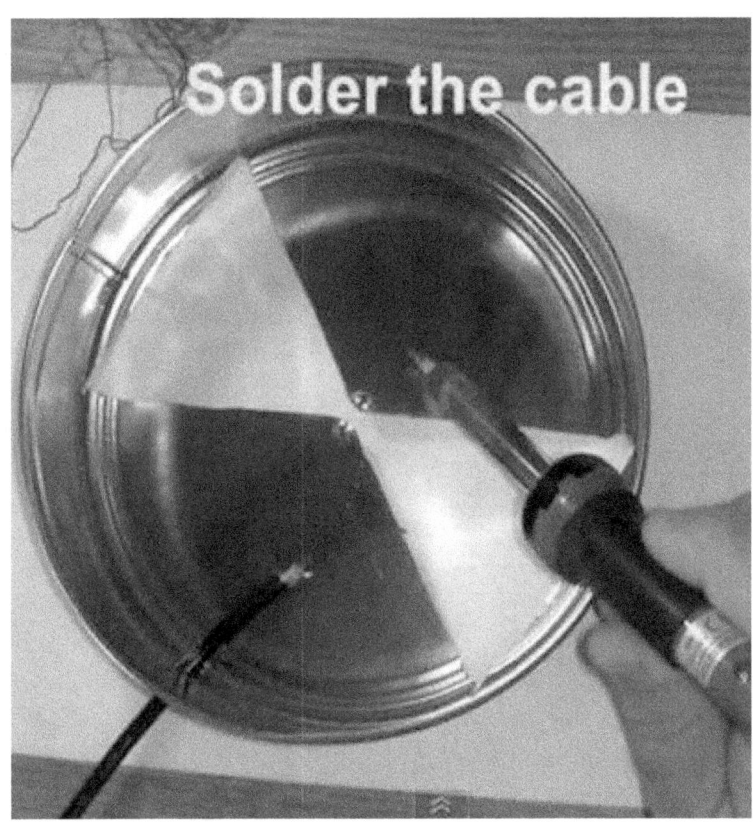

Insert the coax through the ¼ inch hole and then solder it across the two sections in the middle of the tin.

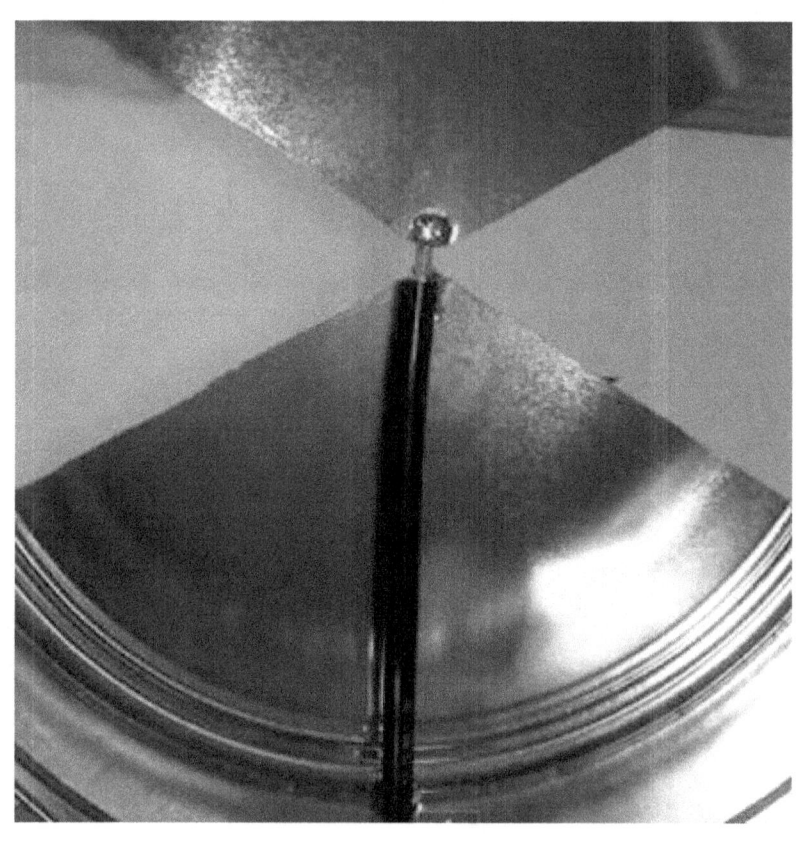

Image shows the soldered cable across the two
sections of cookie tin (the antenna)

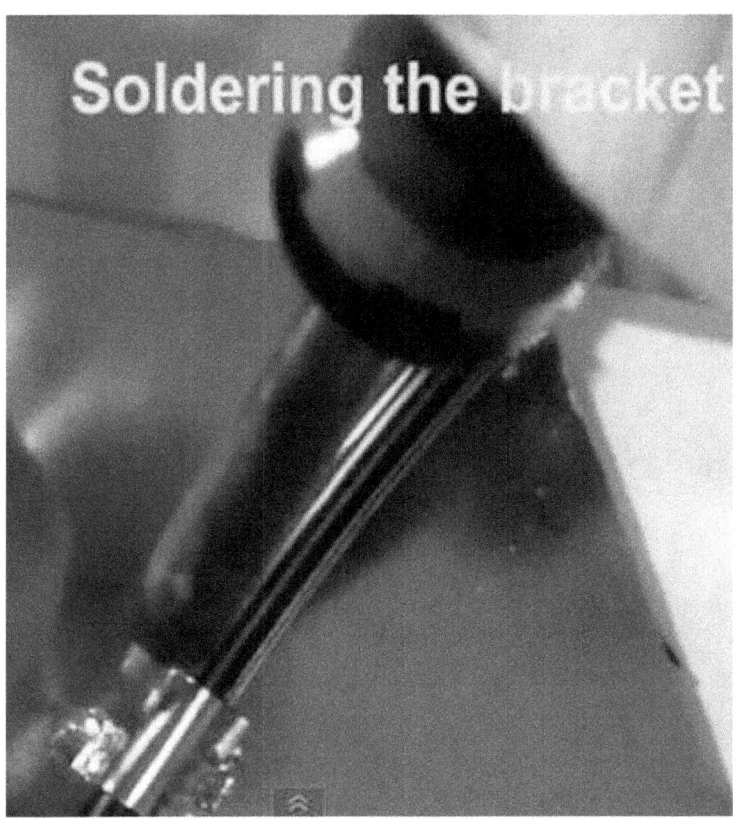

Soldering a bracket to secure the coax to the face of the cookie tin. You can also use a wire or strong tape for this step.

Cut the antenna to about ½ inch in height as in the above image.

Attach double sided tape to the desired window location.

Completed 'cookie-tin' style Antenna

Attach the completed antenna to the window by pressing onto the double sided tape and plug the other end of the coax into your Television. You're done! Remember for any similar antenna design to plug the coax into the ANT 1 IN connection and then press the 'antenna' button on your remote. Run the 'setup' function on your remote and the TV will search through the available channels in your area and add them to the available channel list so they will show up when you are scrolling through the channels with the up or down buttons.

Thank you for choosing this book! I hope it is helpful for you! Please leave a positive review and send a gift copy or lend a gift copy to your friends.

Best regards!—Ryan Seager